ENERGY SECTOR STANDARD OF THE PEOPLE'S REPUBLIC OF CHINA

中华人民共和国能源行业标准

Standard for Ecological Mapping of Hydropower Projects

水电工程生态制图标准

NB/T 10226-2019

Chief Development Department: China Renewable Energy Engineering Institute
Approval Department: National Energy Administration of the People's Republic of China
Implementation Date: May 1, 2020

China Water & Power Press
中国水利水电出版社
Beijing 2024

All rights reserved. No part of this publication may be reproduced, stored in a retrieval system, or transmitted in any form or by any means—electronic, mechanical, photocopying, recording or otherwise, without prior written permission of the publisher.

图书在版编目（CIP）数据

水电工程生态制图标准：NB/T 10226-2019 = Standard for Ecological Mapping of Hydropower Projects（NB/T 10226-2019）：英文 / 国家能源局发布. -- 北京：中国水利水电出版社，2024.7. -- ISBN 978-7-5226-2659-8

Ⅰ．TV222.1-65

中国国家版本馆CIP数据核字第2024MJ4274号

ENERGY SECTOR STANDARD
OF THE PEOPLE'S REPUBLIC OF CHINA
中华人民共和国能源行业标准

Standard for Ecological Mapping of Hydropower Projects
水电工程生态制图标准
NB/T 10226-2019
（英文版）

Issued by National Energy Administration of the People's Republic of China
国家能源局　发布
Translation organized by China Renewable Energy Engineering Institute
水电水利规划设计总院　组织翻译
Published by China Water & Power Press
中国水利水电出版社　出版发行
　　Tel:（+ 86 10）68545888　68545874
　　sales@mwr.gov.cn
　　Account name: China Water & Power Press
　　Address: No.1, Yuyuantan Nanlu, Haidian District, Beijing 100038, China
　　http://www.waterpub.com.cn
中国水利水电出版社微机排版中心　排版
北京中献拓方科技发展有限公司　印刷
184mm×260mm　16开本　2印张　63千字
2024年7月第1版　2024年7月第1次印刷

Price（定价）：￥330.00

Introduction

This English version is one of China's energy sector standard series in English. Its translation was organized by China Renewable Energy Engineering Institute authorized by National Energy Administration of the People's Republic of China in compliance with relevant procedures and stipulations. This English version was issued by National Energy Administration of the People's Republic of China in Announcement [2023] No. 1 dated February 6, 2023.

This version was translated from the Chinese Standard NB/T 10226-2019, *Standard for Ecological Mapping of Hydropower Projects*, published by China Water & Power Press. The copyright is reserved by National Energy Administration of the People's Republic of China. In the event of any discrepancy in the implementation, the Chinese version shall prevail.

Many thanks go to the staff from the relevant standard development organizations and those who have provided generous assistance in the translation and review process.

For further improvement of the English version, any comments and suggestions are welcome and should be addressed to:

China Renewable Energy Engineering Institute
No. 2 Beixiaojie, Liupukang, Xicheng District, Beijing 100120, China
Website: www.creei.cn

Translating organization:

POWERCHINA Zhongnan Engineering Corporation Limited

Translating staff:

| LIU Xiaofen | GUO Yulan | LI Qian | CHEN Lei |

YANG Hong

Review panel members:

LI Zhongjie	POWERCHINA Northwest Engineering Corporation Limited
YU Weiqi	China Renewable Energy Engineering Institute
QIE Chunsheng	Senior English Translator
ZHU Jianzhong	Hohai Univeristy
QI Wen	POWERCHINA Beijing Engineering Corporation Limited

ZHANG Qian	POWERCHINA Guiyang Engineering Corporation Limited
LI Qian	POWERCHINA Chengdu Engineering Corporation Limited
LI Shisheng	China Renewable Energy Engineering Institute

National Energy Administration of the People's Republic of China

翻译出版说明

本译本为国家能源局委托水电水利规划设计总院按照有关程序和规定，统一组织翻译的能源行业标准英文版系列译本之一。2023年2月6日，国家能源局以2023年第1号公告予以公布。

本译本是根据中国水利水电出版社出版的《水电工程生态制图标准》NB/T 10226—2019 翻译的，著作权归国家能源局所有。在使用过程中，如出现异议，以中文版为准。

本译本在翻译和审核过程中，本标准编制单位及编制组有关成员给予了积极协助。

为不断提高本译本的质量，欢迎使用者提出意见和建议，并反馈给水电水利规划设计总院。

  地址：北京市西城区六铺炕北小街2号
  邮编：100120
  网址：www.creei.cn

本译本翻译单位：中国电建集团中南勘测设计研究院有限公司

本译本翻译人员：刘小芬 郭玉兰 李 倩 陈 蕾
       杨 虹

本译本审核人员：

  李仲杰 中国电建集团西北勘测设计研究院有限公司
  喻卫奇 水电水利规划设计总院
  郄春生 英语高级翻译
  祝建中 河海大学
  齐 文 中国电建集团北京勘测设计研究院有限公司
  张 倩 中国电建集团贵阳勘测设计研究院有限公司
  李 茜 中国电建集团成都勘测设计研究院有限公司
  李仕胜 水电水利规划设计总院

国家能源局

Announcement of the National Energy Administration of the People's Republic of China [2019] No. 6

National Energy Administration of the People's Republic of China has approved and issued 384 energy sector standards including *Technical Specification for Electrical Exploration of Hydropower Projects* (Attachment 1), the English version of 48 energy sector standards including *Technical Guide for Rock-Filled Concrete Dams* (Attachment 2), and Amendment Notification No. 1 for 7 energy sector standards including *Technical Code for Environmental Impact Assessment of Wind Farm Projects* (Attachment 3); and abolished 5 energy sector standards/plans including *Charging Standards for Investigation and Design of Wind Power Project* (Attachment 4).

Attachments: 1. Directory of Sector Standards

2. Directory of English Version of Sector Standards

3. Amendment Notification for Sector Standards

4. Directory of Abolished Sector Standards/Plans

National Energy Administration of the People's Republic of China

November 4, 2019

Attachment 1:

Directory of Sector Standards

Serial number	Standard No.	Title	Replaced standard No.	Adopted international standard No.	Approval date	Implementation date
...						
3	NB/T 10226-2019	Standard for Ecological Mapping of Hydropower Projects			2019-11-04	2020-05-01
...						

Foreword

According to the requirements of Document GNKJ [2015] No. 283 issued by National Energy Administration of the People's Republic of China, "Notice on Releasing the Development and Revision Plan of the Energy Sector Standards in 2015", and after extensive investigation and research, summarization of practical experience, consultation of relevant standards, and wide solicitation of opinions, the drafting group has prepared this standard.

The main technical contents of this standard include: basic requirements, base maps, terrestrial ecosystem maps, aquatic ecosystem maps, and legends and labels.

National Energy Administration of the People's Republic of China is in charge of the administration of this standard. China Renewable Energy Engineering Institute has proposed this standard and is responsible for its routine management. Energy Sector Standardization Technical Committee on Hydropower Planning, Resettlement and Environmental Protection is responsible for the explanation of specific technical contents. Comments and suggestions in the implementation of this standard should be addressed to:

China Renewable Energy Engineering Institute
No. 2 Beixiaojie, Liupukang, Xicheng District, Beijing 100120, China

Chief development organizations:

China Renewable Energy Engineering Institute

POWERCHINA Zhongnan Engineering Corporation Limited

Wuhan Imagination Science and Technology Development Co., Ltd.

Participating development organizations:

POWERCHINA Huadong Engineering Corporation Limited

Institute of Hydroecology, MWR & CAS

Chief drafting staff:

YAN Jianbo	MIAO Wenjie	LU Bo	CHEN Jinsheng
YANG Wenzheng	YIN Shuangyu	DAI Xiangrong	LIU Su
LI Xiang	HUANG Daoming	ZHAO Xinchang	CHU Kaifeng
LIU Shengxiang	WANG Jing	ZHAO Kun	WANG Ze
ZHANG Dejian	XIAO Xiaoyun	QIU Jinsheng	YING Feng

JIANG Hao   CAO Yuanyuan   YU Weiqi

Review panel members:

WAN Wengong	XUE Lianfang	CUI Lei	CHEN Yuying
YANG Hongbin	XING Wei	PENG Xin	CHEN Bangfu
LI Jinghua	LIU Qing	YANG Shuhua	CHEN Daqing
ZHANG Jiabo	JIN Yi	NIU Tianxiang	JIANG Hong
ZHANG Rong	LI Jian	QIU Xingchun	PIAO Ling
LI Shisheng			

Contents

1	**General Provisions**	1
2	**Basic Requirements**	2
3	**Base Maps**	3
3.1	General Requirements	3
3.2	Map Types	3
3.3	Map Content	3
4	**Terrestrial Ecosystem Maps**	5
4.1	General Requirements	5
4.2	Map Types	5
4.3	Map Content	5
5	**Aquatic Ecosystem Maps**	7
5.1	General Requirements	7
5.2	Map Types	7
5.3	Map Content	7
6	**Legends and Labels**	9
6.1	Legends	9
6.2	Labels	9
Appendix A	Sizes and Scales of Base Maps	10
Appendix B	Elements of Base Maps	11
Appendix C	Sizes and Scales of Terrestrial Ecosystem Maps	12
Appendix D	Elements of Terrestrial Ecosystem Maps	13
Appendix E	Sizes and Scales of Aquatic Ecosystem Maps	14
Appendix F	Elements of Aquatic Ecosystem Maps	15
Appendix G	Legends for Ecological Elements of Hydropower Projects	16
Explanation of Wording in This Standard		18
List of Quoted Standards		19

1 General Provisions

1.0.1 This standard is formulated with a view to specifying the work base, map content and methods and unifying the technical requirements for the ecological mapping of hydropower projects.

1.0.2 This standard is applicable to the ecological mapping of hydropower projects.

1.0.3 The ecosystem maps of a hydropower project shall be based on the hydropower project information, and display in a systematic manner the spatial relationship between the project and the surrounding eco-environmental elements, the influence characteristics, and the mitigation measures against negative ecological impacts.

1.0.4 The ecosystem maps of a hydropower project shall accurately convey the planning and design intent. The map layout shall be compact, coordinated and explicit, and highlight the theme, with distinct lines, legible lettering, and specified legends and labels.

1.0.5 In addition to this standard, the ecological mapping of hydropower projects shall comply with other current relevant standards of China.

2 Basic Requirements

2.0.1 Ecosystem maps shall consist of base maps, terrestrial ecosystem maps and aquatic ecosystem maps. The mapping work shall be performed in a sequence of data preparation, map making, and final mapping.

2.0.2 The basic data for ecological mapping shall be collected and organized according to the project characteristics and eco-environmental status. If the available data cannot meet the ecological mapping requirements in terms of range, content or accuracy, field investigation, surveying and mapping, remote sensing interpretation, etc. shall be employed to supplement the data.

2.0.3 The elements of an ecosystem map shall consist of the basic elements, project elements and ecological elements. In map making, appropriate map elements shall be selected according to the design intent, to produce an ecosystem map by overlay of elements. The selection of map elements shall meet the following requirements:

1. The base maps shall be developed by overlaying the basic elements with project elements, and shall meet the basic requirements for developing terrestrial ecosystem maps and aquatic ecosystem maps.

2. The terrestrial ecosystem maps and aquatic ecosystem maps shall be developed by overlaying the basic elements and project elements with relevant ecological elements.

2.0.4 In final mapping, appropriate sheet size and map scale and reasonable layout and colors shall be selected according to the mapping purpose. The final map shall explicitly display the map elements, and comprehensively reflect the design intent. A scale bar should be used.

2.0.5 The design of the sheet size, title block, scale, lettering, lines and north arrow for an ecosystem map shall comply with the current sector standard DL/T 5347, *Drawing Standard for Base of Hydropower and Water Conservancy Project*.

2.0.6 When different line types or graphical symbols are used to represent the map elements, the map legend shall be placed at an appropriate position on the map.

3 Base Maps

3.1 General Requirements

3.1.1 The base maps shall clearly and accurately display the geographic location, topography, geomorphy, and eco-environmental features of the region where the project is located.

3.1.2 The sizes and scales of base maps shall be selected in accordance with Appendix A of this standard.

3.2 Map Types

The base maps shall consist of the geographic location map, topographic map, hydrographic map, remote sensing map, current land use map, and distribution map of ecologically sensitive areas.

3.3 Map Content

3.3.1 The elements of base maps shall comply with Appendix B of this standard.

3.3.2 The geographic location map shall clearly display the relative location of the project in the region or riverbasin, and be labeled with the geographic coordinates of the center of the main works.

3.3.3 The topographic map shall display, by contour or digital elevation model, the topographic and geomorphic features of the region where the project is located, and shall be developed according to the following requirements:

1. The scale of the topographic map represented with contours shall be in accordance with Table 3.3.3. When the scale of available topographic maps cannot meet the requirements, remote sensing data shall be used, or field survey shall be conducted.

Table 3.3.3 Scale of topographic map

Length of assessment area L (km)	$L \leq 5$	$5 < L \leq 50$	$50 < L < 100$	$L \geq 100$
Map scale	$\geq 1:10\ 000$	$1:10\ 000 - 1:50\ 000$	$1:50\ 000 - 1:100\ 000$	$\leq 1:100\ 000$

2. The spatial resolution of the topographic map in the form of a digital elevation model should not be lower than 30 m.

3.3.4 The hydrographic map shall clearly display the hydrographic features distribution of the riverbasin where the project is located, and be labeled by text with the hydrological characteristics of the mainstream and tributaries such as name, drainage area, and mean annual flow.

3.3.5 The remote sensing map shall accurately display the ground features in the region where the project is located, and shall be developed according to the following requirements:

1. The ground surface resolutions for 1 : 200 000, 1 : 100 000, 1 : 50 000, and 1 : 10 000 remote sensing maps shall be at least 30.0 m, 15.0 m, 10.0 m and 2.5 m, respectively. Where important vegetation types or key protected wild animals and plants are concentrated, the ground surface resolution shall be at least 2.5 m.

2. The remote sensing information source shall be described with words and placed at an appropriate position on the map, such as the remote sensing image type, calibration accuracy, orbit number, sensor, resolution, imaging time, and waveband combination.

3.3.6 The current land use map shall clearly display the land use types of the land permanently or temporarily occupied by the project and of the surrounding area potentially affected by the project, and shall be developed according to the following requirements:

1. The current land use map shall be based primarily on the land use data surveyed by the Department of Land Resources, and may use the remote sensing interpretation data when necessary.

2. When the remote sensing interpretation is used to develop the current land use map, the land use shall be classified according to the classification requirements of Level 1 land use as specified in GB/T 21010, *Current Land Use Classification*, and Level 1 land use under the codes 05, 06, 07, 08, 09 and 10 may be combined as construction land. The current land use map with a scale not smaller than 1 : 10 000 shall meet the classification requirements of Level 2 land use.

3.3.7 The distribution map of ecologically sensitive areas shall be developed by overlaying the project elements on the function zoning map published by the competent authority, and shall accurately display the positions of the sensitive areas in relation to the project location. When different ecologically sensitive areas overlap, different boundary line types shall be employed to distinguish them.

4 Terrestrial Ecosystem Maps

4.1 General Requirements

4.1.1 The terrestrial ecosystem maps shall clearly and accurately display the characteristics and status of the terrestrial ecosystem in the assessment area, the relationship between the project and ecological elements, and the protection measures and their layout.

4.1.2 The size and scale of a terrestrial ecosystem map shall be selected in accordance with Appendix C of this standard.

4.2 Map Types

The terrestrial ecosystem maps shall consist of the terrestrial ecosystem survey map; the vegetation type distribution map; the distribution maps of key protected wild animals and plants, and old and notable trees; the animal migration route map; the biomass distribution map; the terrestrial ecosystem monitoring scheme map; and the layout map of terrestrial ecosystem protection measures.

4.3 Map Content

4.3.1 The elements of terrestrial ecosystem maps shall comply with Appendix D of this standard.

4.3.2 The terrestrial ecosystem survey map should use the topographic map as the base map, and shall be labeled with the survey range, quadrats, line transects, and belt transects, and show, in tabular form, the size of quadrat, length of line or belt transect, time of survey, etc.

4.3.3 The vegetation type distribution map shall be developed using remote sensing interpretation combined with aerial photography or field survey, and depict the main vegetation types in the region. The vegetation type distribution map shall be prepared in accordance with the following requirements:

1. The vegetation types shall be classified by vegetation subtype. When the map scale is not smaller than 1 : 10 000, the vegetation types shall be accurate to formation group.

2. When the project area involves special vegetation types, their geographic distributions shall be represented separately.

4.3.4 The distribution maps of key protected wild animals and plants, and old and notable trees should use the vegetation type map as the base map, and shall be prepared in accordance with the following requirements:

1 The distribution map of key protected wild animals shall show, in tabular form, the main information such as scientific name, protection level, and quantity.

2 The distribution map of key protected wild plants shall show, in tabular form, the main information such as scientific name, coordinates, elevation, protection level, and quantity.

3 The distribution map of old and notable trees shall show, in tabular form, the main information such as scientific name, coordinates, elevation, protection level, quantity, and age.

4.3.5 The animal migration route map shall use the vegetation type map as the base map, and illustrate the information on main migratory animal species, such as scientific name, migration season, and migration direction.

4.3.6 The biomass distribution map should use the digital elevation map as the base map, and display the levels of biomass in the assessment area with hue or color gradation according to the per unit area biomass. The base map of the vegetative biomass distribution map should be grayscaled.

4.3.7 The terrestrial ecosystem monitoring scheme map should use the vegetation type map as the base map, and shall be labeled with the monitoring sections and the horizontal and vertical monitoring zones, and show, in tabular form, the length and width of horizontal and vertical monitoring zones and the monitoring items, content, time and frequency.

4.3.8 The layout map of terrestrial ecosystem protection measures should use the topographic map as the base map, and shall be labeled with the information on the protection measures, such as spatial location, materials and main design parameters, and show, in tabular form, the names and material specification of the protection measures.

5 Aquatic Ecosystem Maps

5.1 General Requirements

5.1.1 The aquatic ecosystem maps shall clearly and accurately display the characteristics and status of the aquatic ecosystem in the assessment area, the relationship between the project and ecological elements, and the protection measures and their layout.

5.1.2 The size and scale of an aquatic ecosystem map shall be selected in accordance with Appendix E of this standard.

5.2 Map Types

The aquatic ecosystem maps shall consist of the aquatic ecosystem survey map, the important aquatic organisms distribution map, the important fishes distribution map, the main fish habitats distribution map, the fish migration route map, the aquatic ecosystem monitoring scheme map, and the layout map of aquatic ecosystem protection measures.

5.3 Map Content

5.3.1 The elements of aquatic ecosystem maps shall comply with Appendix F of this standard.

5.3.2 The aquatic ecosystem survey map should use the topographic map as the base map, and shall be labeled with the survey sections and sampling points, and show, in tabular form, the information on the survey sections, such as coordinates, elevation, substrate type, water surface width, water depth, and survey time.

5.3.3 The distribution map of important aquatic organisms shall use the hydrographic map as the base map, and shall be labeled with the distribution range of aquatic organisms, and show, in tabular form, the scientific names of important aquatic organisms, preferably with the amount or quantity.

5.3.4 The distribution map of important fishes shall use the hydrographic map as the base map, and shall be labeled with the distribution range of important fishes, and show, in tabular form, the scientific names and protection levels of the fishes.

5.3.5 The distribution map of main fish habitats shall use the hydrographic map and topographic map as the base map, and shall be labeled with the distribution ranges of feeding, spawning and wintering grounds of fishes, and show, in tabular form, the information on the grounds, such as type, area, depth, riverbed terrain, and substrate type.

5.3.6 The fish migration route map shall use the hydrographic map as the base map, and shall be labeled with the migration route and direction, and show, in tabular form, the scientific names and migration seasons of the migratory fishes.

5.3.7 The aquatic ecosystem monitoring scheme map should use the hydrographic map as the base map, and shall be labeled with the monitoring sections, and show, in tabular form, the monitoring items, content, time and frequency.

5.3.8 The layout map of aquatic ecosystem protection measures should use the topographic map as the base map, and shall be labeled with the information on the protection measures, such as spatial location, materials and main design parameters, and show, in tabular form, the names and material specification of the protection measures.

6 Legends and Labels

6.1 Legends

6.1.1 The legends for basic elements shall comply with the current national standard GB/T 20257, *Cartographic Symbols for National Fundamental Scale Maps*.

6.1.2 The legends for the waveband combinations of remote sensing maps shall comply with the current sector standard LY/T 1954, *Technical Regulations of Remote Sensing Base-Map Producing for Forest Resources Inventory*.

6.1.3 The legends for vegetation types shall comply with the current sector standard LY/T 1821, *Cartographic Symbols for Forestry Maps*.

6.1.4 The legends for other ecological elements than those specified in Articles 6.1.1 to 6.1.3 of this standard shall comply with Appendix G of this standard.

6.2 Labels

6.2.1 The labeling for the elements of ecosystem maps shall meet the following requirements:

1. The labels for basic elements shall comply with the current national standard GB/T 20257, *Cartographic Symbols for National Fundamental Scale Maps*.

2. The labels for project elements shall be placed above the project elements or led out by a leader line to the blank on the map to indicate the names of the project elements. The leader line and lettering font of a label shall comply with the current sector standard DL/T 5347, *Drawing Standard for Base of Hydropower and Water Conservancy Project*.

3. The ecological elements should be labeled as for the project elements.

6.2.2 Important element information should be depicted or tabulated at the blank on the map. The table frame line and text shall be the same as the label leader line and text.

Appendix A Sizes and Scales of Base Maps

Table A Sizes and scales of base maps

Map title	Size	Scale
Geographic location map	A3, A4	1 : 200 000, 1 : 100 000, 1 : 50 000
Topographic map	A1, A2, A3, A4	1 : 200 000, 1 : 100 000, 1 : 50 000, 1 : 10 000
Hydrographic map	A1, A2, A3, A4	1 : 200 000, 1 : 100 000, 1 : 50 000, 1 : 10 000
Remote sensing map	A1, A2, A3, A4	1 : 200 000, 1 : 100 000, 1 : 50 000, 1 : 10 000
Current land use map	A1, A2, A3, A4	1 : 200 000, 1 : 100 000, 1 : 50 000, 1 : 10 000
Distribution map of ecologically sensitive areas	A1, A2, A3, A4	1 : 200 000, 1 : 100 000, 1 : 50 000, 1 : 10 000

Appendix B Elements of Base Maps

Table B Elements of base maps

Map title	Basic elements	Project elements
Geographic location map	Administrative division, hydrography, and roads	Main structures and reservoir area
Topographic map	Topography and administrative division	Main structures, inundation line, construction areas, and resettlement areas
Hydrographic map	Hydrography, roads, and administrative division	Main structures, inundation line, construction areas, and resettlement areas
Remote sensing map	Remote sensing imagery, administrative division, hydrography, and roads	Main structures, inundation line, construction areas, and resettlement areas
Current land use map	Land type, administrative division, and hydrography	Main structures, inundation line, construction areas, and resettlement areas
Distribution map of ecologically sensitive areas	Ecologically sensitive areas, administrative division, hydrography, and roads	Main structures, inundation line, construction areas, and resettlement areas

Appendix C Sizes and Scales of Terrestrial Ecosystem Maps

Table C Sizes and scales of terrestrial ecosystem maps

Map title	Size	Scale
Terrestrial ecosystem survey map	A3, A4	1 : 100 000, 1 : 50 000, 1 : 10 000
Vegetation type distribution map	A1, A2, A3, A4	1 : 100 000, 1 : 50 000, 1 : 10 000
Distribution maps of key protected wild animals and plants, and old and notable trees	A1, A2, A3, A4	1 : 100 000, 1 : 50 000, 1 : 10 000
Animal migration route map	A3, A4	1 : 100 000, 1 : 50 000, 1 : 10 000
Biomass distribution map	A1, A2, A3, A4	1 : 50 000, 1 : 10 000, 1 : 2 000
Terrestrial ecosystem monitoring scheme map	A3, A4	1 : 100 000, 1 : 50 000, 1 : 10 000
Layout map of terrestrial ecosystem protection measures	A1, A2, A3, A4	1 : 50 000, 1 : 10 000, 1 : 2 000

Appendix D Elements of Terrestrial Ecosystem Maps

Table D Elements of terrestrial ecosystem maps

Map title	Basic elements	Project elements	Ecological elements
Terrestrial ecosystem survey map	Topography, administrative division, hydrography, and roads	Main structures, inundation line, construction areas, and resettlement areas	Vegetation types, quadrats, line transects, and belt transects
Vegetation type distribution map	Administrative division, hydrography, and roads		Vegetation types
Distribution maps of key protected wild animals and plants, and old and notable trees	Administrative division, hydrography, and roads		Vegetation types and protected objects
Animal migration route map	Topography, administrative division, hydrography, and roads		Migratory animals and migration routes
Biomass distribution map	Topography, administrative division, and hydrography		Biomass
Terrestrial ecosystem monitoring scheme map	Topography, administrative division, hydrography, and roads		Vegetation types, quadrats, line transects, and belt transects
Layout map of terrestrial ecosystem protection measures	Topography, administrative division, and hydrography		Vegetation types, protected objects, and protection measures

Appendix E Sizes and Scales of Aquatic Ecosystem Maps

Table E Sizes and scales of aquatic ecosystem maps

Map title	Size	Scale
Aquatic ecosystem survey map	A3, A4	1 : 100 000, 1 : 50 000, 1 : 10 000
Distribution map of important aquatic organisms	A1, A2, A3, A4	1 : 200 000, 1 : 100 000, 1 : 50 000, 1 : 10 000
Distribution map of important fishes		
Distribution map of main fish habitats		
Fish migration route map	A3, A4	1 : 100 000, 1 : 50 000, 1:10 000
Aquatic ecosystem monitoring scheme map		
Layout map of aquatic ecosystem protection measures	A1, A2, A3, A4	1 : 50 000, 1 : 10 000, 1 : 2 000

Appendix F   Elements of Aquatic Ecosystem Maps

Table F   Elements of aquatic ecosystem maps

Map title	Basic elements	Project elements	Ecological elements
Aquatic ecosystem survey map	Topography, administrative division, and hydrography	Main structures, inundation lines, and construction areas	Survey sections and main habitats
Distribution map of important aquatic organisms	Administrative division and hydrography		Aquatic organisms and fishes
Distribution map of important fishes	Administrative division and hydrography		Fishes
Distribution map of main fish habitats	Topography, administrative division, and hydrography		Fishes, main habitats, and substrate
Fish migration route map	Administrative division and hydrography		Migratory fishes and migration routes
Aquatic ecosystem monitoring scheme map	Topography, administrative division, and hydrography		Monitoring sections and main habitats
Layout map of aquatic ecosystem protection measures	Topography, administrative division, and hydrography		Aquatic organisms, fishes, main habitats, and protection measures

Appendix G   Legends for Ecological Elements of Hydropower Projects

Table G   Legends for ecological elements of hydropower projects

Symbol	Referent	Remarks
(plant in circle, 1)	Key protected wild plant	The number in the symbol may be changed to represent a different key protected wild plant
(tree in circle, 1)	Old and/or notable tree	The number in the symbol may be changed to represent a different old and/or notable tree
(animal in circle, 1)	Key protected wild animal	The number in the symbol may be changed to represent a different key protected wild animal
(fish, 1)	Important fish species	The number in the symbol may be changed to represent a different fish species
①	Sampling, surveying or monitoring point for terrestrial plant	The number in the symbol may be changed to represent a different surveying or monitoring point
◇1	Sampling, surveying or monitoring point for terrestrial animal	The number in the symbol may be changed to represent a different surveying or monitoring point
—①—	Sampling, surveying or monitoring line or belt transect for terrestrial plant	The number in the symbol may be changed to represent a different line or belt transect
—◇1—	Sampling, surveying or monitoring line or belt transect for terrestrial animal	The number in the symbol may be changed to represent a different line or belt transect
⇒1⇒	Animal or fish migration route	The number in the symbol may be changed to represent a different migration route
△1	Sampling, surveying or monitoring point for aquatic organisms	The number in the symbol may be changed to represent a different surveying or monitoring point
—△1—	Sampling, surveying or monitoring section for aquatic organisms	The number in the symbol may be changed to represent a different surveying or monitoring section

Table G *(continued)*

Symbol	Referent	Remarks
	Spawning ground	–
	Feeding ground	–
	Wintering ground	–

Explanation of Wording in This Standard

1. Words used for different degrees of strictness are explained as follows in order to mark the differences in executing the requirements in this standard.

    1) Words denoting a very strict or mandatory requirement:

        "Must" is used for affirmation; "must not" for negation.

    2) Words denoting a strict requirement under normal conditions:

        "Shall" is used for affirmation; "shall not" for negation.

    3) Words denoting a permission of a slight choice or an indication of the most suitable choice when conditions permit:

        "Should" is used for affirmation; "should not" for negation.

    4) "May" is used to express the option available, sometimes with the conditional permit.

2. "Shall meet the requirements of…" or "shall comply with…" is used in this standard to indicate that it is necessary to comply with the requirements stipulated in other relative standards and codes.

List of Quoted Standards

GB/T 20257, *Cartographic Symbols for National Fundamental Scale Maps*

GB/T 21010, *Current Land Use Classification*

DL/T 5347, *Drawing Standard for Base of Hydropower and Water Conservancy Project*

LY/T 1821, *Cartographic Symbols for Forestry Maps*

LY/T 1954, *Technical Regulations of Remote Sensing Base-Map Producing for Forest Resources Inventory*